中国海洋国土空间规划
对外推广科技术语

Terminology of China Marine
Spatial Planning

中国海洋发展基金会海洋空间规划技术重点实验室
国家海洋技术中心　编

海洋出版社

2024 年 · 北京

图书在版编目(CIP)数据

中国海洋国土空间规划对外推广科技术语 / 中国海洋发展基金会海洋空间规划技术重点实验室，国家海洋技术中心编. -- 北京：海洋出版社，2024.6. -- ISBN 978-7-5210-1317-7

Ⅰ. P74

中国国家版本馆 CIP 数据核字第 2024U0M926 号

责任编辑：刘建时
责任印制：安　淼

海洋出版社　出版发行

http://www.oceanpress.com.cn

北京市海淀区大慧寺路 8 号　邮编：100081

北京博海升彩色印刷有限公司印刷　新华书店经销

2024 年 6 月第 1 版　2024 年 6 月北京第 1 次印刷

开本：850 mm×1168 mm　1/32　印张：4.625

字数：208 千字　定价：198.00 元

发行部：010-62100090　总编室：010-62100034

海洋版图书印、装错误可随时退换

中国海洋空间规划对外推广系列公共产品

编　委　会

Committee

LV Bin

PAN Xinchun   **PENG Wei**

WU Shanshan   **TENG Xin**

《中国海洋国土空间规划对外推广科技术语》
编 写 人 员

赵奇威　康琬超　孟　雪

张盼盼　汪　敏　武　文

李　艳　王智能　刘治帅

桑新春　邸梦雅

Authors

ZHAO Qiwei KANG Wanchao MENG Xue

ZHANG Panpan WANG Min   WU Wen

LI Yan   WANG Zhineng LIU Zhishuai

SANG Xinchun DI Mengya

编 者 按

目前，世界上有超过 100 个国家和地区正在开展海洋空间规划相关工作，海洋空间规划已成为世界公认的平衡生态环境保护和海洋经济发展的管理工具，是实现联合国可持续发展目标 14 的有效手段，也正成为全球海洋治理问题的解决途径。

我国海洋国土空间规划经过 40 余年的发展与演变，形成了较为成熟的制度体系、层级体系和分区分类体系。目前我国海洋功能区划和海洋主体功能区规划正在融入国土空间规划体系，继承并不断衍生出各类新时期的海洋空间规划科技词汇。随着我国综合国力的不断提升，海洋国土空间规划技术的不断进步，海洋发展经验的不断积累，我国海洋国土空间规划的理念、技术、方案和经验需要走向国际舞台，与世界沿海国家共享，我国海洋国土空间规划对外推广话语体系亟需构建。海洋国土空间规划科技词汇是构建话语体系的基础，也是我国海洋国土空间规划对外推广的依据。

在本书的编写过程中，为了尽可能将海洋功能区划、海洋主体功能区规划、海岸带规划等海洋规划相关科技语收入书中，我们参考了领导人讲话、中央地方文件、部委报告、权威媒体报道、规划原文、编制指南、技术规范、科学论文等大量文献，将科技语分层分类呈现，包括我国海洋国土空间规划背景、理念、规划类别、规划实施和规划技术（指标）五个层面科技术语的中英文对照释义。同时，我们跟踪梳理了世界典型国家和区域海洋空间规划相关政策，编入本书第六部分，便于开展中国海洋国土空间规划与世界其他国家和地区海洋空间规划相关政策制度的对比和研究，供感兴趣的读者参考。

为了方便读者理解和使用，我们除给出译名外，为每个术语添加了内涵解析。通常情况下，不同的文献对同一个术语会有不同的解读，考虑到本书的使用受众和编写目的，我们尽量选取了最为官方和权威的出处并给出参考文献来源。

本书的编写期望能够为我国海洋国土空间规划理念、技术和实践的对外交流合作提供参考，帮助我国海洋国土空间规划工作者讲好中国海洋故事，为积极参与全球海洋治理，助推"一带一路"规则标准"软联通"提供技术支撑。

由于时间紧迫，加之编者水平所限，本书会有遗漏、欠妥甚至错误之处。欢迎读者在使用过程中提出修订、补充意见，以便今后再版时更加完善。

编　者

2024 年 6 月

目　录

一、背　景

1

二、理 念

三、规划类别

四、规划实施

五、规划技术（指标）

六、国际海洋空间规划

Part 1

一

背景

术语1：人类命运共同体

Terminology 1：A Human Community with a Shared Future

内 涵

人类命运共同体，就是每个民族、每个国家的前途命运都紧紧联系在一起，应该风雨同舟，荣辱与共，努力把我们生于斯、长于斯的这个星球建成一个和睦的大家庭，把世界各国人民对美好生活的向往变成现实[1]。

CONNOTATION

A Human Community with a Shared Future [2] , means that the future and destiny of every nation and country are closely linked. We are in the same boat through thick and thin, we should stick together, share weal and woe, endeavor to build this planet where we were born and raised into a big harmonious family, and turn people's longing for a better life into reality.

参考文献／References

［1］新华社．习近平在中国共产党与世界政党高层对话会上的主旨讲话［EB/OL］.（2017-12-01）［2022-09-30］. http://news. cnr. cn/native/gd/20171201/t20171201_524047053. shtml.

［2］习近平. On Building a Human Community with a Shared Future［M］. 北京：中央编译出版社，2019：1-569.

术语 2：全球发展倡议

Terminology 2：Global Development Initiative（GDI）

内　涵

全球发展倡议包含六条核心理念和原则：坚持发展优先，坚持以人民为中心，坚持普惠包容，坚持创新驱动，坚持人与自然和谐共生，坚持行动导向[1]。

CONNOTATION

Global Development Initiative（GDI）contains six core concepts and principles：Staying committed to development as a priority；Staying committed to a people-centered approach；Staying committed to benefits for all；Staying committed to innovation-driven development；Staying committed to harmony between man and nature；Staying committed to results-oriented actions[2].

参考文献/References

[1] 新华网．习近平在第七十六届联合国大会一般性辩论上的讲话（全文）[EB/OL]．（2021-09-22）[2022-09-21]．www.qstheory.cn/yaowen/2021-09/22/c_1127887219.htm

[2] 新华网．习近平在第七十六届联合国大会一般性辩论上的讲话（双语全文）[EB/OL]．（2021-09-22）[2022-09-21]．https://language.chinadaily.com.cn/a/202109/22/WS614a82b1a310cdd39bc6a948.html

术语 3：海洋命运共同体

Terminology 3: A Maritime Community with a Shared Future

内　涵

海洋命运共同体理念，是对人类命运共同体理念的丰富和发展，是人类命运共同体理念在海洋领域的具体实践[1]。我们人类居住的这个蓝色星球，不是被海洋分割成了各个孤岛，而是被海洋连结成了命运共同体，各国人民安危与共[2]。

CONNOTATION

A Maritime Community with a Shared Future, enriches and develops the concept of "a Human Community with a Shared Future", it is the concrete practice in the marine field of the concept. This blue planet that we share does not consist of isolated islands separated by oceans, but is one connected by oceans as a community with a shared future where people of all countries have a common stake.

参考文献/References

［1］新华每日电讯. 共同构建海洋命运共同体［EB/OL］.（2019-04-24）［2022-09-07］. www. qstheory. cn/llwx/2019-04/24/c_1124407805. htm.

［2］新华社. 习近平集体会见出席海军成立 70 周年多国海军活动外方代表团团长［EB/OL］.（2019-04-23）［2022-09-07］. www. qstheory. cn/yaowen/2019-04/23/c_1124404165. htm.

术语 4：蓝色伙伴关系

Terminology 4：Blue Partnership

内　涵

蓝色伙伴关系是当今中国立足自身发展经验，积极与各国和国际组织在海洋领域构建的，具有开放包容、具体务实、互利共赢特点的伙伴关系[1,2]。

CONNOTATION

Blue Partnership is an open，inclusive，concrete and mutually beneficial partnership that China is actively building with other countries and international organizations in the maritime field based on its own development experience.

参考文献/References

[1] 中华人民共和国中央人民政府. 国家海洋局：倡议有关各方共同建立蓝色伙伴关系［EB/OL］.（2017-04-17）［2022-09-07］. https://www. gov. cn/xinwen/2017-04/17/content_5186532. htm.

[2] 中华人民共和国自然资源部. 2017 厦门国际海洋周开幕［EB/OL］.（2017-11-06）［2022-09-07］. https://mnr. gov. cn/dt/hy/201711/t20171106_2333348. html.

术语 5:"一带一路"倡议

Terminology 5:The Belt and Road Initiative(BRI)

内 涵

"一带一路"倡议是中国国家主席习近平于 2013 年 9 月和 10 月出访中亚和东南亚国家期间先后提出的"丝绸之路经济带"和"21 世纪海上丝绸之路"重大倡议的简称[1]。倡议旨在促进各国互联互通,建立全方位、多层次的复合型互联互通网络,实现沿线国家多元、独立、平衡、可持续发展[2]。

CONNOTATION

The Belt and Road Initiative refers to the initiative of jointly building the Silk Road Economic Belt and the 21st-Century Maritime Silk Road, that were raised by Chinese President Xi Jinping when he visited Central Asia and Southeast Asia in September and October of 2013[3]. The BRI aims to (i) promote the connectivity of countries; (ii) set up all-dimensional, multi-tiered and composite connectivity networks; and (iii) realize diversified, independent, balanced and sustainable development in countries along its pathway[4].

参考文献/References

[1] 中华人民共和国国家发展和改革委员会,中华人民共和国外交部,

中华人民共和国商务部. 推动共建丝绸之路经济带和 21 世纪海上丝绸之路的愿景与行动［EB/OL］.（2015-03-28）［2021-09-18］. https://www. yidaiyilu. gov. cn/yw/qwfb/604. htm.

［2］联合国经济和社会事务部. 携手合作，共享美好未来——"一带一路"倡议支持联合国 2030 年可持续发展议程的进展报告［EB/OL］.（2022-09-19）［2022-10-18］. https://www. yidaiyilu. gov. cn/wcm. files/upload/CMSydylgw/202209/202209190442043. pdf.

［3］National Development and Reform Commission, Ministry of Foreign Affairs, and Ministry of Commerce of the People's Republic of China. Vision and Actions on Jointly Building Silk Road Economic Belt and 21st-Century Maritime Silk Road［EB/OL］.（2015-03-28）［2021-09-18］. https://eng. yidaiyilu. gov. cn/qwyw/qwfb/1084. htm.

［4］Partnering for a Brighter Shared Future-Progress Report on the Belt and Road Initiative in Support of the United Nations 2030 Agenda for Sustainable Development［EB/OL］.（2022-09-19）［2022-10-18］. https://eng. yidaiyilu. gov. cn/wcm. files/upload/CMSydylyw/202209/202209190500015. pdf.

术语 6：政策沟通、设施联通、贸易畅通、资金融通、民心相通

Terminology 6：Policy Coordination，Facilities Connectivity，Unimpeded Trade，Financial Integration，People-to-People Bond

内　涵

"政策沟通""设施联通""贸易畅通""资金融通"和"民心相通"合称"五通"，是"一带一路"倡议的合作重点[1]。

政策沟通是共建"一带一路"的重要保障，是形成携手共建行动的重要先导[2]。加强政府间合作，积极构建多层次政府间宏观政策沟通交流机制，深化利益融合，促进政治互信，达成合作新共识[3]。

设施联通是共建"一带一路"的优先方向[2]。在尊重相关国家主权和安全关切的基础上，沿线国家宜加强基础设施建设规划、技术标准体系的对接，共同推进国际骨干通道建设，逐步形成连接亚洲各次区域以及亚欧非之间的基础设施网络[3]。

贸易畅通是共建"一带一路"的重要内容[2]。宜着力研究解决投资贸易便利化问题，消除投资和贸易壁垒，构建区域内和各国良好的营商环境，积极同沿线国家和地区共同商建自由贸易区，激发释放合作潜力，做大做好合作"蛋糕"，进一步提升沿线各国参与经济全球化的广度和深度[2,3]。

9

资金融通是共建"一带一路"的重要支撑[2]。深化金融合作，推进亚洲货币稳定体系、投融资体系和信用体系建设。扩大沿线国家双边本币互换、结算的范围和规模。推动亚洲债券市场的开放和发展。共同推进亚洲基础设施投资银行、金砖国家开发银行筹建，有关各方就建立上海合作组织融资机构开展磋商。加快丝路基金组建运营[3]。

民心相通是共建"一带一路"的人文基础[2]。传承和弘扬丝绸之路友好合作精神，各国广泛开展文化交流、学术往来、人才交流合作、媒体合作、青年和妇女交往、志愿者服务等，为深化双多边合作奠定坚实的民意基础[3]。

CONNOTATION

Policy Coordination, Facilities Connectivity, Unimpeded Trade, Financial Integration, People-to-People Bond are the cooperation priorities of the Belt and Road Initiative, which are referred to as Five Pillars[4].

Policy Coordination is an important guarantee for the Belt and Road Initiative, and an essential precondition for joint actions[5]. We should promote intergovernmental cooperation, build a multi-level intergovernmental macro policy exchange and communication mechanism, expand shared interests, enhance mutual political trust and reach new cooperation consensus[6].

Facilities Connectivity is a priority area for the Belt and Road Initiative[5]. On the basis of respecting each other's sovereignty and security concerns, countries along the Belt and Road should improve the connectivity of their infrastructure construction plans and technical standard systems, jointly push forward the construction of international trunk passageways, and form an infrastructure network connecting all sub-regions in Asia, and between Asia, Europe and Africa step by step[6].

Unimpeded Trade is an important goal of the Belt and Road Initiative[5]. We should strive to improve investment and trade facilitation, and remove investment and trade barriers for the creation of a sound business environment within the region and in all related countries. We will discuss with countries and regions along the Belt and Road on opening free trade areas so as to unleash the potential for expanded cooperation and enable the participants to engage in broader and deeper economic globalization[5,6].

Financial Integration is an important pillar of the Belt and Road Initiative[5]. We should deepen financial cooperation, and make more efforts in building a currency stability system, investment and financing system and credit information system in Asia. We should expand the scope and scale of bilateral currency swap and settlement with other countries along the Belt and Road, open and develop the bond market in Asia, make joint efforts to establish the Asian Infrastructure Investment Bank and BRICS New Development Bank, conduct negotiation among related parties on establishing Shanghai Cooperation Organization (SCO) financing institution, and set up and put into operation the Silk Road Fund as early as possible[6].

People-to-People Bond are the cultural foundation for building the Belt and Road[5]. We should carry forward the spirit of friendly cooperation of the Silk Road by promoting extensive cultural and academic exchanges, personnel exchanges and cooperation, media cooperation, youth and women exchanges and volunteer services, so as to win public support for deepening bilateral and multilateral cooperation[6].

参考文献/References

[1] 联合国经济和社会事务部. 携手合作, 共享美好未来——"一带一路"倡议支持联合国 2030 年可持续发展议程的进展报告[EB/OL]. (2022-09-19)[2022-10-18]. https://www. yidaiyilu. gov. cn/wcm. files/upload/CMSydylgw/202209/202209190442043. pdf.

[2] 推进"一带一路"建设工作领导小组办公室. 共建"一带一路"倡议: 进展、贡献与展望[EB/OL]. (2019-04-22)[2022-09-26]. https://www. yidaiyilu. gov. cn/wcm. files/upload/CMSydylgw/201904/201904220250016. pdf.

[3] 中华人民共和国国家发展和改革委员会, 中华人民共和国外交部, 中华人民共和国商务部. 推动共建丝绸之路经济带和 21 世纪海上丝绸之路的愿景与行动[EB/OL]. (2015-03-28)[2021-09-18]. https://www. yidaiyilu. gov. cn/yw/qwfb/604. htm.

[4] Partnering for a Brighter Shared Future-Progress Report on the Belt and Road Initiative in Support of the United Nations 2030 Agenda for Sustainable Development [EB/OL]. (2022-09-19)[2022-10-18]. https://eng. yidaiyilu. gov. cn/wcm. files/upload/CMSydylyw/202209/202209190500015. pdf.

[5] Office of the Leading Group for Promoting the Belt and Road Initiative. The Belt and Road Initiative Progress, Contributions andProspects[EB/OL]. (2019-04-22)[2022-09-26]. https://eng. yidaiyilu. gov. cn/zchj/qwfb/86739. htm.

[6] National Development and Reform Commission, Ministry of Foreign Affairs, and Ministry of Commerce of the People's Republic of China. Vision and Actions on Jointly Building Silk Road Economic Belt and 21st-Century Maritime Silk Road [EB/OL]. (2015-03-28) [2021-09-18]. https://eng. yidaiyilu. gov. cn/qwyw/qwfb/1084. htm.

术语 7："双循环"新发展格局

Terminology 7："Dual Circulation" New Development Pattern

内　涵

"双循环"新发展格局指的是以国内大循环为主体、国内国际双循环相互促进的新发展格局。构建新发展格局的关键在于经济循环的畅通无阻，最本质的特征是实现高水平的自立自强[1]。

CONNOTATION

The "Dual Circulation" New Development Pattern refers to a new development pattern whereby domestic and foreign markets can boost each other, with the domestic market as the mainstay. The key to building the "Dual Circulation" New Development Pattern lies in the unimpeded economic circulation and the most essential feature is to achieve a high level of self-reliance.

参考文献/References

[1] 中华人民共和国国家发展和改革委员会规划司．"十四五"规划《纲要》解读文章之 10｜推动形成以国内大循环为主体、国内国际双循环相互促进的新发展格局［EB/OL］.（2021-12-25）［2022-09-27］. https://www.ndrc.gov.cn/fggz/fzzlgh/gjfzgh/202112/t20211225_1309698.html.

Part 2

二

理念

术语1：生态文明

Terminology 1：Ecological Civilization
（Eco-Civilization）

内　涵

　　生态文明是人类社会进步的重大成果，是实现人与自然和谐共生的必然要求。建设生态文明，要以资源环境承载能力为基础，以自然规律为准则，以可持续发展、人与自然和谐为目标，坚定走生产发展、生活富裕、生态良好的文明发展道路，建设美丽中国[1]。

CONNOTATION

　　Ecological civilization（Eco-Civilization）[2] is a major achievement of human social progress and a necessary requirement for harmonious coexistence between human and nature. Building Ecological Civilization should take the carrying capacity of resources and environment as the basis, follow the laws of nature, aim for sustainable development and harmony between human and nature, pursue a civilized development path that ensures increased levels of production, prosperous lives, and a sound ecology, and build a Beautiful China.

参考文献／References

[1] 中共中央宣传部．习近平新时代中国特色社会主义思想三十讲[M]．

北京：学习出版社，2018：242-243.

［2］中央编译局中央文献重要术语译文审定委员会．中央文献重要术语译文发布(第八期)［EB/OL］．(2015-12-10)［2022-08-16］．www.scio.gov.cn/zhzc/35353/35354/Document/1507114/1507114.htm.

术语 2：绿水青山就是金山银山

Terminology 2：Lucid waters and lush mountains are invaluable assets

内 涵

绿水青山就是金山银山，阐述了经济发展和生态环境保护的关系。良好的生态环境是最公平的公共产品、最普惠的民生福祉。绿水青山既是自然财富、生态财富，又是社会财富、经济财富[1]。

CONNOTATION

Lucid waters and lush mountains are invaluable assets[2], represents the relationship between economic development and ecological environmental protection. A good ecological environment is the fairest public product and the most accessible welfare for the people [3]. Lucid waters and lush mountains are not only natural and ecological wealth, but also social and economic wealth.

参考文献／References

[1] 中共中央宣传部．习近平新时代中国特色社会主义思想学习纲要 [M]．北京：学习出版社 人民出版社，2019：169-171．

[2] 中央编译局中央文献重要术语译文审定委员会．中央文献重要术语译文发布（第八期）[EB/OL]．（2015-12-10）[2022-08-16]．www. scio. gov. cn/zhzc/35353/35354/Document/1507114/1507114. htm.

［3］Chinadaily. Lucid waters and lush mountains are invaluable assets［EB/
OL］.（2017-10-09）［2021-11-16］. http：//www. chinadaily. com. cn/
china/19thcpcnationalcongress/2017-10/09/content_33032118. htm.

术语3：山水林田湖草沙冰一体化保护和系统治理

Terminology 3: Integrated Protection and Systematic Governance of Mountains, Rivers, Forests, Farmlands, Lakes, Grasslands, Sand and Ice

内　涵

山水林田湖草沙冰一体化保护和系统治理是保护和系统治理的新格局，强调重要生态系统整体性保护，统筹治山治水治城，遵循生态系统的整体性、系统性及其内在规律，开展山水林田湖草沙冰共治。山水林田湖草沙冰是一个生命共同体[1,2]。

CONNOTATION

Integrated protection and systematic governance of mountains, rivers, forests, farmlands, lakes, grasslands, sand and ice is a new pattern of protection and systematic governance, emphasizing the overall protection of important ecosystems, governing mountains, waters and cities coordinately, and following the integrity, systematization and inherent laws of the ecosystems, so as to carry out co-governance of mountains, rivers, forests, farmlands, lakes, grass-

21

lands, sand and ice. Mountains, rivers, forests, farmlands, lakes, grasslands, sand and ice are a life community.

参考文献/References

[1] 新华社. 关于《中共中央关于全面深化改革若干重大问题的决定》的说明[EB/OL]. (2013-11-18)[2021-11-16]. http://www.moa.gov.cn/ztzl/qjqh/zyjs/201311/t20131119_3679950.htm.

[2] 青海省人民政府. 全力打造山水林田湖草沙冰保护和系统治理新高地[EB/OL]. (2021-10-27)[2022-09-30]. http://www.qinghai.gov.cn/zwgk/system/2021/10/27/010395662.shtml.

术语 4：自然资源"两统一"

Terminology 4："Two Unification" of Natural Resources

内　涵

自然资源"两统一"，指统一行使全民所有自然资源资产所有者职责，统一行使所有国土空间用途管制和生态保护修复职责[1]。

CONNOTATION

"Two Unification" of Natural Resources means that, in a unified way, perform the duties of the owner of public-owned natural resource assets and the duties of regulating the use of all territorial space and protecting and restoring ecosystems [2].

参考文献/References

［1］新华社. 习近平：决胜全面建成小康社会夺取新时代中国特色社会主义伟大胜利——在中国共产党第十九次全国代表大会上的报告［EB/OL］.（2017-10-27）［2021-05-16］. http://www. 12371. cn/2017/10/27/ARTI1509103656574313. shtml.

［2］新华社. Full text of Xi Jinping's report at 19th CPC National Congress ［EB/OL］.（2017-11-03）［2021-05-16］. https://www. xinhuanet. com/english/special/2017-11/03/c_136725942. htm.

术语5：多规合一

Terminology 5：Integrate Plans into Single Master Plan

内　涵

多规合一是指将主体功能区规划、土地利用规划、城乡规划等空间规划融合为统一的国土空间规划，建立国土空间规划体系并监督实施，以强化国土空间规划对各专项规划的指导约束作用[1]。

CONNOTATION

Integrate Plans into Single Master Plan[2] means to integrate spatial planning, including main functional areas planning, land use planning, urban and rural planning into a unified territorial spatial planning, establish, supervise the implementation of the territorial spatial planning system, so as to strengthen the guiding and constraining role of the territorial spatial planning in various special plannings.

参考文献/References

[1] 中华人民共和国中央人民政府. 中共中央 国务院关于建立国土空间规划体系并监督实施的若干意见 [EB/OL]. （2019-05-23）[2021-11-02]. https://www.gov.cn/gongbao/content/2019/content_5397679.htm.

［2］ Report on the Work of the Government Delivered at the Fifth Session of the 12th National People's Congress of the People's Republic of China on March 5, 2017 ［EB/OL］. （2017-03-17） ［2021-09-02］. http://language. chinadaily. com. cn/2017-03/17/content_28591593. htm.

术语6：陆海统筹

Terminology 6: Coordinated Land and Marine Development

内　涵

陆海统筹是指从我国陆海兼备的国情出发，在进一步优化提升陆域国土开发的基础上，以提升海洋在国家发展全局中的战略地位为前提，加强海洋在资源环境保障、经济发展和国家安全维护中的作用，通过海陆资源开发、交通通道建设、生态保护等领域的统筹协调，促进海陆两大系统的优势互补、良性互动和协调发展，增强国家对海洋的管控与利用能力，建设海洋强国，构建大陆文明与海洋文明相容并济的可持续发展格局[1]。

CONNOTATION

Coordinated Land and Marine Development[2] means that based on the national conditions of both land and sea, on the basis of further optimizing and improving land development, on the premise of enhancing the strategic position of the oceans in the overall development of the country, strengthen the role of the oceans in resources and environment guarantee, economic development and national security maintenance, promote complementarity, positive interaction and coordinated development of sea and land systems through coordinating the development of land and sea resources, transport channel construction and ecological protection, strengthen the country's capacity to regulate and utilize the sea, build a strong maritime country; and build a sustainable development pattern in which both continental and marine civilizations are compatible.

参考文献/References

［1］央视网. 童心向党读热词：陆海统筹［EB/OL］.（2018-03-22）［2021-05-16］. http://shaoer. cctv. com/2018/03/22/ARTIdhq41I6fp3AF0z Fsanv9180322. shtml.

［2］新华社. Full text of Xi Jinping's report at 19th CPC National Congress ［EB/OL］.（2017-11-03）［2021-05-16］. https://www. xinhuanet. com/ english/special/2017-11/03/c_136725942. htm.

Part 3

三

规划类别

术语 1：国土空间规划

Terminology 1：Territorial Spatial Planning

内 涵

国土空间规划是对一定区域国土空间开发保护在空间和时间上作出的安排，包括总体规划、详细规划和相关专项规划[1]。

CONNOTATION

Territorial Spatial Planning refers to the spatial and temporal arrangement for the development and protection of a certain region's territorial space，including Overall Planning，Detailed Planning and Special Planning.

参考文献/References

[1] 中华人民共和国中央人民政府. 中共中央 国务院关于建立国土空间规划体系并监督实施的若干意见[EB/OL]. (2019-05-23)[2022-10-02]. http://www.gov.cn/gongbao/content/2019/content_5397679.htm.

术语 2：总体规划

Terminology 2：Overall Planning

内　涵

　　总体规划是对国土空间保护、开发、利用、修复作出的总体部署与统筹安排，是各类开发保护建设活动的基本依据，也是详细规划的依据、相关专项规划的基础。市县及以上编制总体规划[1]。

CONNOTATION

　　Overall Planning is the overall and coordinated arrangements for the protection, development, utilization and restoration, is the basis for various development, protection and construction activities, and for the Detailed Plannings and relevant Special Plannings. The Overall Planning is formulated at municipal and county level and above.

参考文献/References

[1] 中华人民共和国中央人民政府. 中共中央 国务院关于建立国土空间规划体系并监督实施的若干意见[EB/OL]. (2019-05-23)[2022-10-02]. http://www.gov.cn/gongbao/content/2019/content_5397679.htm.

术语 3：详细规划

Terminology 3：Detailed Planning

内　涵

　　详细规划是对具体地块用途和开发建设强度等作出的实施性安排，是开展国土空间开发保护活动、实施国土空间用途管制、核发城乡建设项目规划许可、进行各项建设等的法定依据。在市县及以下编制详细规划[1]。

CONNOTATION

Detailed Planning is the practical arrangement for uses and development and construction intensity of specific land blocks，is the legal basis for the development and protection of territorial spaces，the use regulation of territorial spaces，the issuance of planning permissions for urban and rural construction projects，and the various constructions. The Detailed Planning is formulated at municipal and county level and below.

参考文献/References

[1] 中华人民共和国中央人民政府. 中共中央 国务院关于建立国土空间规划体系并监督实施的若干意见[EB/OL]. (2019-05-23)[2022-10-02]. http://www.gov.cn/gongbao/content/2019/content_5397679.htm.

术语 4：专项规划

Terminology 4：Special Planning

内　涵

专项规划是指在特定区域(流域)、特定领域，为体现特定功能，对空间开发保护利用作出的专门安排，是涉及空间利用的专项规划[1]。

CONNOTATION

Special Planning refers to the special arrangements for spatial development, protection and utilization in specific regions (river basins) and specific fields to reflect specific functions. It is the special planning referring to spatial utilization.

参考文献/References

[1] 中华人民共和国中央人民政府. 中共中央 国务院关于建立国土空间规划体系并监督实施的若干意见[EB/OL]. (2019-05-23) [2022-10-02]. http://www.gov.cn/gongbao/content/2019/content_5397679.htm.

术语5：海岸带综合保护与利用规划

Terminology 5：Integrated Coastal Zone Protection and Utilization Planning （ICZP）

内　涵

海岸带综合保护与利用规划是国土空间规划的专项规划，是陆海统筹的专门安排，是海岸带高质量发展的空间蓝图，是国土空间规划在海岸带区域针对特定问题的细化、深化和补充。规划为海岸带地区资源保护与利用、生态保护与修复、灾害防御等提供管理依据，为海岸带产业与滨海人居环境布局优化提供空间指引，为海岸带地区实施用途管制提供基础[1,2,4]。

CONNOTATION

The Integrated Coastal Zone Protection and Utilization Planning （ICZP） is one of the Special Planning under Territorial Spatial Planning, is the special arrangements to coordinate land and sea, is the spatial blueprint for the high-quality development of coastal zone, is the refinement, deepening and supplement of Territorial Spatial Planning for specific issues in coastal zone. ICZP provides management basis for the resources protection and utilization, ecological protection and restoration, disaster prevention and mitigation in coastal zone, provides spatial guidance for optimizing the layout of coastal industries and coastal human settlements, provides basis for the use regulation in coastal zone.

一级类 Zones

术语 5-1：生态保护区

Terminology 5-1：Ecological Protection Zone

内　涵

具有特殊重要生态功能或生态敏感脆弱、必须强制性严格保护的陆地和海洋自然区域，包括陆域生态保护红线、海洋生态保护红线集中划定的区域[3,4]。

CONNOTATION

The terrestrial and marine natural areas with special important ecological functions or sensitive and fragile ecology that must be strictly protected, including the areas designated by the terrestrial ecological conservation redline and marine ecological conservation redline.

术语 5-2：生态控制区

Terminology 5-2：Ecological Control Zone

内　涵

生态保护红线外，需要予以保留原貌、强化生态保育和生态建设、限制开发建设的陆地和海洋自然区域[3,4]。

CONNOTATION

The terrestrial and marine natural areas outside the ecological conservation redline, which original appearance needs to be retained, ecological conservation and ecological construction need to be strengthen, and development and construction need to be restricted.

术语 5-3：海洋发展区

Terminology 5-3：Marine Development Zone

内　涵

允许集中开展开发利用活动的海域，以及允许适度开展开发利用活动的无居民海岛，包括渔业用海区、交通运输用海区、工矿通信用海区、游憩用海区、特殊用海区、海洋预留区[3,4]。

CONNOTATION

The sea areas where centralized development and utilization activities are allowed, and uninhabited islands that allow moderate development and utilization activities, including Marine Fishery Zone, Marine Transportation Zone, Marine Industry Mining and Communication Zone, Marine Tourism and Recreation Zone, Marine Special Purpose Zone and Marine Reservation Zone.

二级类 Subzones

术语 5-3-1：渔业用海区

Terminology 5-3-1：Marine Fishery Zone

内　涵

以渔业基础设施建设、养殖和捕捞生产等渔业利用为主要功能导向的海域和无居民海岛[3,4]。

CONNOTATION

The sea areas and uninhabited islands with the main functions of fishery infrastructure construction, aquaculture and fishing production.

术语 5-3-2：交通运输用海区

Terminology 5-3-2：Marine Transportation Zone

内　涵

以港口建设、路桥建设、航运等为主要功能导向的海域和无居民海岛[3,4]。

CONNOTATION

The sea areas and uninhabited islands with the main functions of port construction, road and bridge construction and shipping.

术语 5-3-3：工矿通信用海区

Terminology 5-3-3: Marine Industry Mining and Communication Zone

内　涵

以临海工业利用、矿产能源开发和海底工程建设等为主要功能导向的海域和无居民海岛[3,4]。

CONNOTATION

The sea areas and uninhabited islands with the main functions of coastal industrial utilization, mineral energy development and subsea engineering construction.

术语 5-3-4：游憩用海区

Terminology 5-3-4: Marine Tourism and Recreation Zone

内　涵

以开发利用旅游资源为主要功能导向的海域和无居民海岛[3,4]。

CONNOTATION

The sea areas and uninhabited islands with the main functions of the development and utilization of tourism resources.

术语5-3-5：特殊用海区

Terminology 5-3-5：Marine Special Purpose Zone

内　涵

以污水达标排放、倾倒、军事等特殊利用为主要功能导向的海域和无居民海岛[3,4]。

CONNOTATION

The sea areas and uninhabited islands with the main functions of standard sewage discharge, dumping, military and other special uses.

术语5-3-6：海洋预留区

Terminology 5-3-6：Marine Reservation Zone

内　涵

规划期内为重大项目用海用岛预留的控制性后备发展区域[3,4]。

CONNOTATION

The controlled reserve development areas reserved for major projects that need to use sea areas and/or islands during the planning period.

参考文献/References

［1］中华人民共和国中央人民政府. 中共中央 国务院关于建立国土空间规划体系并监督实施的若干意见［EB/OL］.（2019-05-23）［2022-10-02］. http://www. gov. cn/gongbao/content/2019/content_5397679. htm.

［2］中国海洋信息网. 自然资源部 2019 年全国两会建议提案办结，79 件涉海——推动海洋领域高质量发展［EB/OL］.（2020-05-21）［2021-12-22］. http://www. nmdis. org. cn/c/2020-05-21/71634. shtml.

［3］中华人民共和国自然资源部. 市级国土空间总体规划编制指南（试行）［EB/OL］.（2020-09-22）［2022-10-02］. http://gi. mnr. gov. cn/202009/t20200924_2561550. html.

［4］中华人民共和国自然资源部. 省级海岸带综合保护与利用规划编制指南（试行）［EB/OL］.（2020-09-22）［2022-10-02］. http://gi. mnr. gov. cn/202109/t20210913_2680305. html.

术语 6: 海洋功能区划

Terminology 6: Marine Functional Zoning

内 涵

海洋功能区划是指根据海域的区位条件、自然环境、自然资源、开发保护现状和经济社会发展的需要，按照海洋功能标准，将海域划分为不同使用类型和不同环境质量要求的功能区，用以控制和引导海域的使用方向，保护和改善海洋生态环境，促进海洋资源的可持续利用[1]。

CONNOTATION

Marine Functional Zoning refers to the division of sea areas into functional zones with different use types and different environmental quality requirements according to the marine functional standards, based on the location conditions, natural environment, natural resources, development & conservation status and economic & social development demand of sea areas. It is used to regulate and guide the use of sea areas, protect and improve the marine ecological environment, and promote the sustainable use of marine resources.

一级类 Zones

术语 6-1：农渔业区

Terminology 6-1：Agriculture and Fishery Zone

内 涵

农渔业区是指适于拓展农业发展空间和开发海洋生物资源，可供农业围垦，渔港和育苗场等渔业基础设施建设，海水增养殖和捕捞生产，以及重要渔业品种养护的海域，包括农业围垦区、渔业基础设施区、养殖区、增殖区、捕捞区和水产种质资源保护区(重要渔业品种养护区)[2]。

CONNOTATION

Agriculture and Fishery Zone[3] refers to the sea areas suitable for expanding the space for mariculture development and developing marine biological resources, as well as for agriculture reclamation, fishery infrastructure construction such as fishing ports and breeding grounds, mariculture, nursery and capturing production, and the conservation of important fishery species, including Agriculture Reclamation Zone, Fishery Infrastructure Zone, Mariculture Zone, Nursery Zone, Capture Zone and Conservation Zone of Key Fishery Species.

二级类 Subzones

术语 6-1-1：农业围垦区

Terminology 6-1-1：Agriculture Reclamation Zone

内　涵

农业围垦区是指供围垦后用于农、林、牧业生产的海域[4]。

CONNOTATION

Agriculture Reclamation Zone refers to the sea areas used for agriculture, forestry and animal husbandry production after reclamation.

术语 6-1-2：渔业基础设施区

Terminology 6-1-2：Fishery Infrastructure Zone

内　涵

渔业基础设施区是指适于渔业基础设施建设，供渔船停靠、进行装卸作业和避风的海域，以及用来繁殖重要苗种的场所[4]。

CONNOTATION

Fishery Infrastructure Zone[3] refers to the sea areas suitable for the construction of fishery infrastructure, for the docking of fishing boats, loading and unloading operations and shelter from the wind, as well as the place used for breeding important seedling.

术语 6-1-3：养殖区

Terminology 6-1-3: Mariculture Zone

内　涵

养殖区是指供养殖或培育海洋经济动物和植物的海域[4]。

CONNOTATION

Mariculture Zone[3] refers to the sea areas for breeding or cultivating marine economic animals and plants.

术语 6-1-4：增殖区

Terminology 6-1-4: Nursery Zone

内　涵

增殖区是指经过繁殖保护措施来增加和补充生物群体数量的海域[4]。

CONNOTATION

Nursery Zone[3] refers to the sea areas where reproductive protection measures are used to increase and replenish the number of biological populations.

术语 6-1-5：捕捞区

Terminology 6-1-5：Capture Zone

内　涵

捕捞区是指在海洋游泳生物(鱼类和大型无脊椎动物)产卵场、索饵场、越冬场及其洄游通道(即过路渔场)使用国家规定的渔具或人工垂钓的方式获取海产经济动物的海域[4]。

CONNOTATION

Capture Zone[3] refers to the sea areas in which marine swimming organisms (fish and large invertebrates) spawning grounds, feeding grounds, wintering grounds and migration channels use fishing gear prescribed by the State or artificial fishing to obtain marine economic animals.

术语 6-1-6：水产种质资源保护区(重要渔业品种养护区)

Terminology 6-1-6：Conservation Zone of Key Fishery Species

内　涵

水产种质资源保护区(重要渔业品种养护区)是指用来保护具有重要经济价值、遗传育种价值以及重要科研价值的渔业品种及其产卵场、索饵场、越冬场和洄游路线等栖息繁衍生境的海域[4]。

CONNOTATION

Conservation Zone of Key Fishery Species[3] refers to the sea areas used to protect fishery species with important economic value, genetic breeding value and scientific research value and their habitats such as spawning grounds, feeding grounds, wintering grounds and migration routes.

一级类 Zones

术语6-2: 港口航运区

Terminology 6-2: Port and Navigation Zone

内 涵

港口航运区是指适于开发利用港口航运资源, 可供港口、航道和锚地建设的海域, 包括港口区、航道区和锚地区[2]。

CONNOTATION

Port and Navigation Zone[3] refers to the sea areas suitable for the development and utilization of port and navigation resources, and available for the construction of ports, waterways and anchorages, including Port Zone, Waterway Zone and Anchorage Zone.

二级类 Subzones

术语 6-2-1：港口区

Terminology 6-2-1：Port Zone

内　涵

港口区是指供船舶停靠、进行装卸作业、避风及货物存放的海域，包括港口内港池、码头和仓储地等[4]。

CONNOTATION

Port Zone[3] refers to the sea areas for ships docking, loading and unloading operations, taking shelter from wind and storing goods, including harbor pools, wharves and warehouse land.

术语 6-2-2：航道区

Terminology 6-2-2：Waterway Zone

内　涵

航道区是指供船只航行使用的海域[4]。

CONNOTATION

Waterway Zone[3] refers to the sea areas for ships navigation.

术语 6-2-3：锚地区

Terminology 6-2-3：Anchorage Zone

内　涵

锚地区是指供船舶候潮、待泊、联检、避风以及进行水上装卸作业的海域[4]。

CONNOTATION

Anchorage Zone[3] refers to the sea areas for ships to wait for tide and berth, joint inspection and shelter from wind, and carry out water loading and unloading operations.

一级类 Zones

术语 6-3：工业与城镇用海区

Terminology 6-3：Industry and Urban Use Zone

内　涵

工业与城镇用海区是指适于发展临海工业与滨海城镇的海域，包括工业用海区和城镇用海区[2]。

CONNOTATION

The Industrial and Urban Use Zone[3] refers to the sea areas suitable for developing coastal industries and coastal urban, including Industrial Construction Zone and Urban Construction Zone.

二级类 Subzones

术语 6-3-1：工业用海区（工业建设区）

Terminology 6-3-1：Industrial Construction Zone

内　涵

工业用海区（工业建设区）是指供临海企业和工业园区建设的海域[4]。

CONNOTATION

Industrial Construction Zone[3] refers to the sea areas for the construction of coastal enterprises and industrial parks.

术语 6-3-2：城镇用海区（城镇建设区）

Terminology 6-3-2：Urban Construction Zone

内　涵

城镇用海区（城镇建设区）是指供沿海市政设施、滨海新城和海上机场等建设的海域[4]。

CONNOTATION

Urban Construction Zone[3] refers to the sea areas for the construction of coastal municipal facilities, coastal urbans and maritime airports.

一级类 Zones

术语 6-4：矿产与能源区

Terminology 6-4：Minerals and Energy Zone

内　涵

矿产与能源区是指适于开发利用矿产资源与海上能源，可供油气和固体矿产等勘探、开采作业，以及盐田和可再生能源等开发利用的海域，包括油气区、固体矿产区、盐田区和可再生能源区[2]。

CONNOTATION

Minerals and Energy Zone[3] refers to the sea areas suitable for the exploitation and utilization of mineral resources and marine energy, for the exploration and mining operations of oil and gas and solid minerals, and for the exploitation and utilization of salt pans and renewable energy, including Oil and Gas Zone, Solid Mineral Zone, Salt Pan and Renewable Energy Zone.

二级类 Subzones

术语 6-4-1：油气区

Terminology 6-4-1：Oil and Gas Zone

内　涵

油气区是指供油气勘探和开采作业的海域[4]。

CONNOTATION

Oil and Gas Zone[3] refers to the sea areas for the exploration and mining operations of oil and gas.

术语6-4-2：固体矿产区

Terminology 6-4-2：Solid Mineral Zone

内　涵

固体矿产区指供固体矿产勘探和开采作业的海域[4]。

CONNOTATION

Solid Mineral Zone[3] refers to the sea areas for the exploration and mining operations of solid mineral.

术语6-4-3：盐田区

Terminology 6-4-3：Salt Pan

内　涵

盐田区是指供养水、制卤和晒盐等盐业生产的海域[4]。

CONNOTATION

Salt Pan[3] refers to the sea areas for the salt industry production including water providing, brine making and bay salt.

术语 6-4-4：可再生能源区

Terminology 6-4-4：Renewable Energy Zone

内　涵

可再生能源区是指供开发利用潮汐能、波浪能等可再生能源的海域[4]。

CONNOTATION

Renewable Energy Zone[3] refers to the sea areas for the exploitation and utilization of renewable energy such as tidal energy and wave energy.

一级类 Zones

术语 6-5：旅游休闲娱乐区

Terminology 6-5：Tourism and Entertainment Zone

内　涵

旅游休闲娱乐区是指适于开发利用滨海和海上旅游资源，可供旅游景区开发和海上文体娱乐活动场所建设的海域，包括风景旅游区和文体休闲娱乐区[2]。

CONNOTATION

Tourism and Entertainment Zone[3] refers to the sea areas suit-

53

able for the development and utilization of coastal and marine tourism resources, and available for the development of tourist attractions and the construction of marine recreational, cultural and entertainment places, including Tourism Zone and Recreation Zone.

二级类 Subzones

术语 6-5-1：风景旅游区

Terminology 6-5-1：Tourism Zone

内　涵

风景旅游区是指具有一定质和量的自然景观和人文景观、可供游人参观游览的海域[4]。

CONNOTATION

Tourism Zone[3] refers to the sea areas with a certain quality and quantity of natural and cultural landscapes which are available for visitors sightseeing.

术语 6-5-2：文体休闲娱乐区（文体娱乐区）

Terminology 6-5-2：Recreation Zone

内　涵

文体休闲娱乐区（文体娱乐区）是指供海上文体娱乐、运动和度假等的海域[4]。

CONNOTATION

Recreation Zone[3] refers to the sea areas for taking marine recreational, cultural and entertainment activities, sports and holidays.

一级类 Zones

术语6-6：海洋保护区

Terminology 6-6: Marine Protection Zone

内 涵

海洋保护区是指专供海洋资源、环境和生态保护的海域，包括海洋自然保护区、海洋特别保护区[2]。

CONNOTATION

Marine Protection Zone[3] refers to the sea areas exclusively for the protection of marine resources, environment and ecology, including Marine Protected Area and Special Marine Protected Area.

二级类 Subzones

术语6-6-1：海洋自然保护区

Terminology 6-6-1: Marine Protected Area

内 涵

海洋自然保护区是指为保护珍稀、濒危海洋生物物种和经济生物物种及其栖息地，以及有重大科学、文化、景观和生态服务价值的海洋自然客体、自然生态系统和历史遗迹等需要而划定的海域[4]。

CONNOTATION

Marine Protected Area[3] refers to the sea areas delimited for protecting the rare and endangered marine biological species and economic biological species and their habitats, as well as marine natural objects, natural ecosystems and historical sites of great scientific, cultural, landscape and ecological service values.

术语 6-6-2：海洋特别保护区

Terminology 6-6-2：Special Marine Protected Area

内　涵

海洋特别保护区是指具有特殊地理条件、生态系统、生物与非生物资源，以及海洋开发利用特殊需要的海域[4]。

CONNOTATION

Special Marine Protected Area[3] refers to the sea areas with special geographical conditions, ecosystems, living and non-living resources, and special needs for marine development and utilization.

一级类 Zones

术语 6-7：特殊利用区

Terminology 6-7：Special Purpose Zone

内　涵

特殊利用区是指供其他特殊用途排他使用的海域[2]。

CONNOTATION

Special Purpose Zone[3] refers to the sea areas used exclusively for other special purposes.

二级类 Subzones

术语 6-7-1：军事区

Terminology 6-7-1：Military Zone

内　涵

军事区是指供军事用途排他使用的海域[4]。

CONNOTATION

Military Zone[3] refers to the sea areas used exclusively for military purpose.

术语 6-7-2：其他特殊利用区

Terminology 6-7-2：Other Special Purpose Zones

内 涵

其他特殊利用区是指供海底管线铺设、路桥建设、污水达标排放、倾倒等特殊用途排他使用的海域[4]。

CONNOTATION

Other Special Purpose Zones[3] refer to the sea areas used exclusively for special purposes such as laying of submarine pipelines, construction of roads and bridges, standard sewage discharge and dumping.

一级类 Zones

术语 6-8：保留区

Terminology 6-8：Reservation Zone

内 涵

保留区是指为保留海域后备空间资源，专门划定的在区划期限内限制开发的海域[2]。

CONNOTATION

Reservation Zone[3] refers to the sea areas specially delimited for reserving the space resources, which is limited for development within the period of planning.

参考文献/References

[1] 张宏声. 全国海洋功能区划概要[M]. 北京：海洋出版社，2003：3-4.

[2] 中华人民共和国自然资源部. 国务院批准《全国海洋功能区划（2011—2020 年）》（公开版）[EB/OL]. (2012-04-18) [2022-10-14]. https://f. mnr. gov. cn/201806/t20180621_1830454. html.

[3] TENG X, ZHAO Q, ZHANG P, et al., Implementing marine functional zoning in China. Marine Policy, 2021, 132：103484. https://doi. org/10. 1016/j. marpol. 2019. 02. 055.

[4] 中华人民共和国自然资源部. 省级海洋功能区划编制技术要求. (2010-10-15) [2022-10-14]. http://gc. mnr. gov. cn/201806/t20180614_1795624. html.

术语7：海洋主体功能区规划

Terminology 7：Marine Main Functional Areas Planning

内　涵

　　海洋主体功能区规划是根据不同海域资源环境承载能力、现有开发强度和发展潜力，合理确定不同海域主体功能，构建陆海协调、人海和谐的海洋空间开发格局[1]。

　　海洋主体功能区按开发内容可分为产业与城镇建设、农渔业生产、生态环境服务三种功能。依据主体功能，将海洋空间划分为以下四类区域：优化开发区域、重点开发区域、限制开发区域、禁止开发区域[1]。

CONNOTATION

Marine Main Functional Areas Planning refers to reasonably determining the main functions of different sea areas and establishing a marine space development pattern of land and sea coordination and harmony between human and sea, according to the resources and environment carrying capacity, and the existing development intensity and potentials.

According to the development content, the Marine Main Functional Areas can be divided into three functions: industry and urban construction; agricultural and fishery production; and ecological environment service. Based on the main functions, the marine spaces are delimited into four types: Optimized Development Area, Key Development Area, Limited Development Area and Prohibited Development Area.

术语7-1：优化开发区

Terminology 7-1: Optimized Development Area

内　涵

优化开发区是指现有开发利用强度较高，资源环境约束较强，产业结构亟需调整和优化的海域[1]。

CONNOTATION

Optimized Development Area refers to the sea areas with high intensity of development and utilization, strong constraints on resources and environment, and therefore in urgent need to adjust and optimize the industrial structure.

术语7-2：重点开发区

Terminology 7-2: Key Development Area

内　涵

重点开发区是指在沿海经济社会发展中具有重要地位、发展潜力较大、资源环境承载能力较强、可以进行高强度集中开发的海域[1]。

CONNOTATION

Key Development Area refers to sea areas that play an important role in coastal economic and social development, still have great potential for development and strong resources and environment carrying capacity and therefore can be developed intensively and centralizedly.

术语 7-3：限制开发区

Terminology 7-3: Limited Development Area

内　涵

限制开发区是指以提供海洋水产品为主要功能的海域，包括用于保护海洋渔业资源和海洋生态功能的海域[1]。

CONNOTATION

Limited Development Area refers to the sea areas with the provision of marine aquatic products as the main function, including the sea areas used to protect marine fishery resources and marine ecological functions.

术语 7-4：禁止开发区

Terminology 7-4: Prohibited Development Area

内　涵

禁止开发区是指对维护海洋生物多样性，保护典型海洋生态系统具有重要作用的海域，包括海洋自然保护区、领海基点所在岛屿等[1]。

CONNOTATION

Prohibited Development Area refers to the sea areas that play an important role in maintaining marine biodiversity and protecting typical marine ecosystems, including marine protected areas and the islands where the base points of the territorial sea are located.

参考文献/References

[1] 中国政府网．国务院关于印发全国海洋主体功能区规划的通知 [EB/OL]．（2015-08-20）[2022-09-30]．http://www.gov.cn/zhengce/content/2015-08/20/content_10107.htm.

术语 8：海域使用规划
Terminology 8：Sea Area Use Planning

内　涵

海域使用规划指在一定海域内，根据海洋功能区划以及当地经济和社会发展状况、海洋环境保护和海上交通安全的要求，对海域资源的开发、利用、治理和保护在空间上、时间上所做的科学设计和安排[1,2]。

CONNOTATION

Sea Area Use Planning refers to the scientifically spatial and temporal design and arrangement for the development, utilization, governance and protection of marine resources in certain sea areas, according to Marine Functional Zoning, the development status of local economy and society, and the requirements of marine environment protection and marine traffic safety.

参考文献/References

［1］青岛市海洋发展局. 山东省海域使用管理条例［EB/OL］.（2015-07-27）［2022-10-06］. https://ocean. qingdao. gov. cn/zcfg/shj/202110/t20211030_3702469. shtml.

［2］山东省海洋与渔业厅. 山东省海洋与渔业厅关于印发《山东省县级海域使用规划管理办法（试行）》的通知［EB/OL］.（2013-10-29）［2022-10-06］. http://www. shandong. gov. cn/art/2013/10/29/art_2259_24163. html.

术语 9：区域用海规划

Terminology 9：Regional Sea Area Use Planning

内　涵

区域用海规划是依据全国和省级海洋功能区划，客观分析涉及海域的自然条件及面临形势，明确说明区域用海整体围填的必要性、可行性，提出区域发展的功能定位、空间布局方案和规划期限内年度围填海计划规模，并对规划实施可能产生的环境影响进行全面分析、预测和评估。对于连片开发、需要整体围填用于建设或农业开发的海域，省级海洋行政主管部门要指导市、县级人民政府组织编制区域用海规划。区域用海规划分为区域建设用海规划和区域农业围垦用海规划[1]。

CONNOTATION

Regional Sea Area Use Planning means, in accordance with national and provincial marine functional zoning, to objectively ana-lyse the natural conditions and situations of the sea areas involved, clearly explain the necessity and feasibility of the overall reclamation of regional sea area use, propose the functions of the regional development, the spatial layout plan and the scale of the annual reclamation plan within the planning period, and conduct the comprehensive analysis, forecast and assessment of the possible environmental impacts from the implementation of the Planning. For sea areas that require contiguous development, or need to be reclaimed as a whole for construction or agricultural development, the provincial marine administrative department shall guide the municipal and county-level people's governments to organize the formulation of Regional Sea Area Use Planning. The Regional Sea Area Use Planning is divided into Regional Construction Sea Area Use Planning and Regional Agricultural Reclamation Sea Area Use Planning[1].

参考文献/References

[1] 中华人民共和国自然资源部. 国家发展改革委 国家海洋局关于加强围填海规划计划管理的通知(发改地区〔2009〕2976 号)[EB/OL]. (2010-03-03)[2022-10-06]. http://gc. mnr. gov. cn/201806/t20180614_1795619. html.

术语 10：海岛保护规划

Terminology 10：Island Protection Planning

内　涵

海岛保护规划是引导全社会保护和合理利用海岛资源的纲领性文件，是从事海岛保护、利用活动的依据。海岛保护规划全面分析当前海岛保护与利用的现状、存在的问题和面临的形势，提出规划目标，明确海岛分类、分区保护的具体要求，确定了海岛资源和生态调查评估、偏远海岛开发利用等十项重点工程，并在组织领导、法制建设、能力建设、公众参与、工程管理和资金保障方面提出具体保障措施[1]。

CONNOTATION

Island Protection Planning is a programmatic document that guides the whole society to protect and rationally use island resources, and is the basis for island protection and utilization activities. The Planning comprehensively analyzes the status, existing problems and situations of island protection and utilization, proposes the planning objectives, clarifies the specific requirements for island classification and zoning protection, determines ten key projects including island resources and ecological survey and assessment and remote island development and utilization and puts forwards specific safeguard measures in terms of organization and

67

leadership, legal system building, capacity building, public partici-pation, project management and capital guarantee[1].

参考文献/References

[1] 中华人民共和国自然资源部.《全国海岛保护规划》公布实施[EB/OL].（2012-04-19）[2021-12-22]. http://g. mnr. gov. cn/201701/t20170123_1428738. html.

术语11：专属经济区和大陆架规划

Terminology 11：Exclusive Economic Zone and Continental Shelf Planning （EEZCSP）

内　涵

专属经济区和大陆架规划是专属经济区和大陆架资源保护利用的时间和空间安排，有利于行使国家对专属经济区和大陆架的主权权利和管辖权[1]。

CONNOTATION

Exclusive Economic Zone and Continental Shelf Planning （EEZCSP）refers to the temporal and spatial arrangement for the resources protection and utilization in the exclusive economic zone and continental shelf areas，which is conducive to the state to exercise sovereign rights and jurisdiction over its exclusive economic zone and continental shelf.

参考文献/References

［1］中华人民共和国专属经济区和大陆架法［J］. 中华人民共和国最高人民法院公报，1998（04）：115-116.

术语 12：自然保护地

Terminology 12：Protected Areas（PAs）

内　涵

自然保护地是由各级政府依法划定或确认，对重要的自然生态系统、自然遗迹、自然景观及其所承载的自然资源、生态功能和文化价值实施长期保护的陆域或海域[1]。

CONNOTATION

Protected Areas（PAs）is the land or sea areas delimited or confirmed by governments at all levels to protect important natural ecosystems, natural relics, natural landscapes and the natural resources, ecological functions and cultural values carried by them for a long time.

参考文献/References

[1] 中华人民共和国中央人民政府. 中共中央办公厅 国务院办公厅印发《关于建立以国家公园为主体的自然保护地体系的指导意见》[EB/OL].（2019-06-26）[2022-10-09]. http://www.gov.cn/xinwen/2019-06/26/content_5403497.htm.

Part 4

四

规划实施

术语 1：中华人民共和国海域使用管理法

Terminology 1：Law of the People's Republic of China on Administration of the Use of Sea Areas

内　涵

《中华人民共和国海域使用管理法》是为了加强海域使用管理，维护国家海域所有权和海域使用权人的合法权益，促进海域的合理开发和可持续利用而制定的法律[1]。

CONNOTATION

Law of the People's Republic of China on Administration of the Use of Sea Areas[2] is enacted to strengthen the administration of the use of sea areas, safeguard the state ownership of sea areas and the legitimate rights and interests of the use right owners of sea areas, and promote the rational development and sustainable use of sea areas.

参考文献/References

[1] 中华人民共和国中央人民政府. 中华人民共和国主席令 第六十一号 [EB/OL]. (2001-10-27) [2022-09-23]. http://www.gov.cn/gongbao/content/2001/content_61173.htm.

[2] 中华人民共和国中央人民政府. State Council Gazette Issue No. 34 Serial No. 1033 (Dec 10, 2001) [EB/OL]. (2001-10-27) [2022-11-22]. http://english.www.gov.cn/archive/state_council_gazette/2014/10/29/content_281475002787061.htm

术语2：中华人民共和国海洋环境保护法

Terminology 2：Marine Environment Protection Law of the People's Republic of China

内　涵

《中华人民共和国海洋环境保护法》是为了保护和改善海洋环境，保护海洋资源，防治污染损害，维护生态平衡，保障人体健康，促进经济和社会的可持续发展而制定的法律[1]。

CONNOTATION

Marine Environment Protection Law of the People's Republic of China is enacted to protect and improve the marine environment, conserve marine resources, prevent pollution damages, maintain ecological balance, safeguard human health and promote sustainable economic and social development[2].

参考文献/References

［1］全国人民代表大会．中华人民共和国海洋环境保护法［EB/OL］．（2017-11-28）［2022-09-30］．http：//www.npc.gov.cn/npc/sjxflfg/201906/604accded97d4cc39268d7f16720ef1d.shtml.

［2］中华人民共和国中央人民政府．Marine Environment Protection Law of the People's Republic of China［EB/OL］．（2014-08-23）［2022-09-30］．http：//english.www.gov.cn/archive/press_briefing/2014/08/23/content_281474983042445.htm.

术语 3：中华人民共和国海岛保护法

Terminology 3：Sea Island Protection Law of the People's Republic of China

内　涵

《中华人民共和国海岛保护法》是为了保护海岛及其周边海域生态系统，合理开发利用海岛自然资源，维护国家海洋权益，促进经济社会可持续发展而制定的法律[1]。

CONNOTATION

Sea Island Protection Law of the People's Republic of China is enacted to protect the ecosystem of islands and their surrounding waters, rationally develop and utilize natural resources on islands, safeguard the maritime rights and interests of the state, and promote sustainable development of economy and society.

参考文献/References

[1] 中华人民共和国中央人民政府. 中华人民共和国海岛保护法［EB/OL］.（2009-12-26）［2022-11-22］. http：//www. gov. cn/flfg/2009-12/26/content_1497461. htm.

术语 4：国土空间用途管制

Terminology 4：Territorial Spatial Use Regulation

内 涵

国土空间用途管制是以总体规划、详细规划为依据，对陆海所有国土空间的保护、开发和利用活动，按照规划确定的区域、边界、用途和使用条件等核发行政许可，进行行政审批等过程[1]。

CONNOTATION

Territorial Spatial Use Regulation refers to the process of issuing administrative licenses and conducting administrative examination and approval for the protection, development and utilization activities in all territorial spaces including land and sea areas based on overall plannings and detailed plannings, in accordance with the areas, boundaries, uses and conditions of use determined by the plannings.

参考文献/References

［1］中华人民共和国自然资源部．省级国土空间规划编制指南（试行）
［EB/OL］．（2020-01-17）［2022-10-02］．http://gi.mnr.gov.cn/202001/
P020200120642346540184.pdf.

术语5：自然资源统一确权登记

Terminology 5：Unified Confirmation and Registration of Natural Resource Rights

内　涵

自然资源统一确权登记是指对水流、森林、山岭、草原、荒地、滩涂、海域、无居民海岛以及探明储量的矿产资源等自然资源的所有权和所有自然生态空间统一进行确权登记的制度[1]。

CONNOTATION

Unified Confirmation and Registration of Natural Resource Rights refers to the system of the unified confirmation and registration of the ownership of natural resources such as water flow, forests, mountains, grasslands, barren land, tidal flats, sea areas, uninhabited islands and proved reserves of mineral resources and all natural ecological spaces.

参考文献/References

[1] 中华人民共和国中央人民政府．五部门印发《自然资源统一确权登记暂行办法》[EB/OL]．（2019-07-23）[2021-12-07]．http://www.gov.cn/xinwen/2019-07/23/content_5413117.htm.

术语 6：自然资源资产价值评估

Terminology 6：Valuation of Natural Resource Assets

内　涵

　　自然资源资产的价值包括其资源价值、环境价值、生态价值、经济价值、社会价值和文化价值等在内，其中经济价值是自然资源资产价值的核心体现。自然资源资产价值评估，是正确认识和评价自然资源资产价值的基础性工作，通过成本法、收益法、市场法、意愿法等来评估自然资源资产的价值或价格[1]。

CONNOTATION

The value of natural resource assets includes its resource value, environmental value, ecological value, economic value, social value and cultural value, among which economic value is the core embodiment. The Valuation of Natural Resource Assets is the basic work to correctly understand and evaluate the value of natural resource assets, which evaluates the value or price of natural resource assets through cost method, income method, market method and willingness method.

参考文献／References

[1] 谷树忠，李维明. 自然资源资产价值及其评估[EB/OL]. (2016-01-11)[2022-11-25]. jer. whu. edu. cn/jjgc/15/2016-01-11/2231. html.

术语 7：海域有偿使用制度

Terminology 7: Paid Use of Sea Area System

内　涵

海域有偿使用制度是指单位和个人使用海域，应当按照国务院的规定缴纳海域使用金。国家实行海域有偿使用制度，海域使用金应当按照国务院的规定上缴财政[1]。

CONNOTATION

Paid Use of Sea Area System means that any units and individuals who use sea areas shall pay fees in accordance with the regulations of the State Council. The State implements Paid Use of Sea Area System, the fees shall be turned over to the Treasury in accordance with the regulations of the State Council.

参考文献/References

[1] 中华人民共和国中央人民政府. 中华人民共和国主席令 第六十一号 [EB/OL]. (2001-10-27) [2022-09-23]. http://www. gov. cn/gongbao/content/2001/content_61173. htm.

术语 8：海域使用金

Terminology 8: Sea Area Use Fee

内　涵

海域使用金是对使用国家海洋资源的补偿，是海域有偿使用制度的核心内容，海域使用者使用了国家的海洋资源，向国家缴纳海域使用金是一种应当支付的代价[1]。

CONNOTATION

Sea Area Use Fee is the compensation for the use of national marine resources and the core elements of the Paid Use of Sea Area System. Sea area users use the marine resources of the state, so that the payment of Sea Area Use Fee for the state is also a kind of price that should be paid.

参考文献/References

[1] 青岛市海洋与渔业局.《中华人民共和国海域使用管理法》释义——海域使用管理的法律依据［EB/OL］.（2015-08-16）［2022-09-22］. http://www.qingdao.gov.cn/zwgk/xxgk/hyfz/gkml/zcjd/202010/t20201018_412973.shtml.

术语 9：海域使用申请审批

Terminology 9：Sea Area Use Application and Approval

内　涵

海域使用申请是指单位和个人可以向县级以上人民政府海洋行政主管部门申请使用海域。海域使用审批是指县级以上人民政府海洋行政主管部门依据海洋功能区划，对海域使用申请进行审核，并依照《中华人民共和国海域使用管理法》和省、自治区、直辖市人民政府的规定，报有批准权的人民政府批准。海洋行政主管部门审核海域使用申请，应当征求同级有关部门的意见[1]。

CONNOTATION

Sea Area Use Application means that units and individuals may apply for the use of sea areas to the marine administrative department of the government at or above the county level. Sea Area Use Approval means that the marine administrative department under the government at or above the county level shall examine the sea area use applications on the basis of Marine Functional Zoning, and then submit the applications for approval to the government with approval authority in accordance with *Law of the People's Republic of China on Administration of the Use of Sea Areas* and the regulations of the government at provincial, autonomous regional and municipal levels. When examining the sea area use applications, the marine administrative department shall solicit opinions from related departments at the same level.

参考文献/References

[1] 中华人民共和国中央人民政府. 中华人民共和国主席令 第六十一号 [EB/OL]. (2001-10-27) [2022-09-23]. http://www.gov.cn/gongbao/content/2001/content_61173.htm.

术语 10：海域使用论证

Terminology 10: Sea Area Use
Demonstration

内　涵

海域使用论证是在详细了解和勘查项目所在区域海洋资源生态、开发利用现状和权属状况的基础上，依据生态优先、节约集约原则，科学客观地分析论证项目用海的必要性、选址与规模的合理性、对海洋资源和生态的影响范围与程度、规划符合性和利益相关者的协调性等，提出项目生态用海对策，并给出明确的用海论证结论[1]。使用海域应当依法进行海域使用论证[2]。

CONNOTATION

Sea Area Use Demonstration is to scientifically and objectively analyze and demonstrate the necessity of sea area use of the project, the rationality of site selection and scale, the scope and degree of impact on marine resources and ecology, the conformity of planning and the coordination of stakeholders, etc. , and then put forward the countermeasures of ecological sea area use for the project, and give the clear conclusion of Sea Area Use Demonstration, in accordance with the principles of ecological priority and intensive use, on the basis of detailed understanding and survey of the marine resource ecology, development and utilization status and ownership status of the area where the project is located. Using sea areas should carry out Sea Area Use Demonstration according to law.

参考文献/References

［1］中华人民共和国中央人民政府．自然资源部关于规范海域使用论证材料编制的通知［EB/OL］．（2021-01-08）［2022-11-29］．https://www.gov.cn/zhengce/zhengceku/2021-01/14/content_5579825.htm.

［2］中华人民共和国自然资源部．关于印发《海域使用权管理规定》的通知［EB/OL］．（2006-10-13）［2022-11-29］．https://f.mnr.gov.cn/201807/t20180705_2019289.html.

术语 11：海域使用权

Terminology 11：Sea Area Use Right

内　涵

海域使用权是一种自然资源的使用权，它是指非所有人依照法律规定，为一定的目的使用国家所有的海洋资源[1]。海域属于国家所有，国务院代表国家行使海域所有权，单位和个人使用海域，必须依法取得海域使用权[2]。

CONNOTATION

Sea Area Use Right is a kind of right to use natural resources, it means that the non-owner uses the marine resources owned by the State for a certain purpose according to laws. The sea areas belong to the State, the State Council shall exercise the ownership of the sea areas on behalf of the State. The units and individuals using the sea areas must obtain Sea Area Use Right.

参考文献/References

[1] 青岛市海洋与渔业局.《中华人民共和国海域使用管理法》释义——海域使用管理的法律依据[EB/OL].（2015-08-16）[2022-09-22]. http://www. qingdao. gov. cn/zwgk/xxgk/hyfz/gkml/zcjd/202010/t20201018_412973. shtml.

[2] 中华人民共和国中央人民政府. 中华人民共和国主席令 第六十一号 [EB/OL].（2001-10-27）[2022-09-23]. http://www. gov. cn/gongbao/content/2001/content_61173. htm.

术语 12：集约节约用海

Terminology 12：Intensive Sea Area Use

内　涵

集约节约用海是在海域现有的资源禀赋、社会经济和技术条件下，以可持续发展为前提，兼顾经济、社会和生态环境效益，通过立体利用、标准控制、布局优化、市场配置等手段，达到减少海域过度开发、提高海域资源利用效率的目的[1]。

CONNOTATION

Intensive Sea Area Use is to reduce over-exploitation of sea areas and improve the utilization efficiency of sea area resources by means of three-dimensional utilization，standard control，layout optimization and market allocation under the existing resource endowment，socio-economic and technological conditions，on the premise of sustainable development and taking into account economic，social and ecological environmental benefits.

参考文献/References

［1］王晶. 探索节约集约用海新模式［N］. 中国自然资源报，2021-09-15（005）. DOI：10. 28291/n. cnki. ngtzy. 2021. 003167.

术语 13：亲海空间

Terminology 13：Sea-enjoyable Space

内　涵

亲海空间是具有景观性、功能性、生态性和人文性，具备人和海洋亲近条件的海岸带空间[1]。

CONNOTATION

Sea-enjoyable Space is the coastal zone space with landscape, functionality, ecology, humanity and having the conditions for people to get close to the sea.

参考文献/References

[1] 王冀. 城市滨水区亲水空间场所精神的塑造[D]. 北京：中国林业科学研究院, 2012.

Part 5

五

规划技术（指标）

术语 1：海域使用分类

Terminology 1: Sea Area Use Classification

内　涵

海域使用分类指按照一定的原则，划分海域使用类型并界定其用海方式[1]。

CONNOTATION

Sea Area Use Classification refers to the division of the sea areas use types and the definition of the sea areas use ways according to certain principles.

参考文献/References

[1] 中华人民共和国自然资源部. 关于印发《海域使用分类体系》和《海籍调查规范》的通知[EB/OL]. (2008-05-06)[2022-11-29]. https://gc. mnr. gov. cn/201806/t20180614_1795677. html.

术语 2："双评价"

Terminology 2："Double Assessments"

内　涵

"双评价"由资源环境承载能力评价和国土空间开发适宜性评价两部分构成。

资源环境承载能力是指基于特定发展阶段、经济技术水平、生产生活方式和生态保护目标，一定地域范围内资源环境要素能够支撑农业生产、城镇建设等人类活动的最大合理规模。

国土空间开发适宜性是指在维系生态系统健康和国土安全的前提下，综合考虑资源环境等要素条件，特定国土空间进行农业生产、城镇建设等人类活动的适宜程度[1]。

CONNOTATION

"Double Assessments" consists of Assessment of Resources and Environmental Carrying Capacity and Assessment of the Suitability of Territorial Spatial Development.

Resources and Environmental Carrying Capacity refers to the maximum reasonable scale of human activities that resources and environmental factors can support in a certain region, such as agricultural production and urban construction, based on the specific development stage, economic and technological level, production and life style and ecological protection target.

The Suitability of Territorial Spatial Development refers to the suitability of carrying out human activities, such as agricultural production and urban construction, in a specific territorial space, under the premise of maintaining ecosystem health and national security, and taking resources and environment and other factors into comprehensive considerations.

参考文献/References

[1] 中华人民共和国中央人民政府. 自然资源部办公厅关于印发《资源环境承载能力和国土空间开发适宜性评价指南（试行）》的函[EB/OL]. (2020-01-19)［2022-11-17］. http://www. gov. cn/zhengce/zhengceku/2020-01/22/content_5471523. htm.

术语 3："双评估"

Terminology 3："Double Evaluations"

内 涵

"双评估"是指国土空间开发保护现状评估和现行空间类规划实施情况评估。

国土空间开发保护现状评估一般以安全、创新、协调、绿色、开放、共享等理念构建的指标体系为标准，从数量、质量、布局、结构、效率等角度，找出一定区域国土空间开发保护现状与高质量发展要求之间存在的差距和问题所在。同时可在现状评估的基础上，结合影响国土空间开发保护因素的变动趋势，分析国土空间发展面临的潜在风险。

现行空间类规划实施情况评估（规划评估）指对现行土地利用总体规划、城乡总体规划、林业草业规划、海洋功能区划等空间类规划，在规划目标、规模结构、保护利用等方面的实施情况进行评估，并识别不同空间规划之间的冲突和矛盾，总结成效和问题[1]。

CONNOTATION

"Double Evaluations" refers to Evaluation of the Development and Protection Status of Territorial Space and Evaluation of the Implementation of Existing Spatial Plannings.

Evaluation of the Development and Protection Status of Territo-

rial Space (Status Evaluation) is to find the gaps and problems between the status of development & protection and the high-quality
requirements in a certain region from the perspectives of quantity,
quality, layout, structure and efficiency, with a standard of the in-
dex system based on the ideas of safety, innovation, coordination,
green, open and sharing. At the same time, on the basis of Status
Evaluation, the potential risks of territorial spatial development can
be analysed combined with the changing trend of the factors
affecting the development and protection of territorial space.

Evaluation of the Implementation of Existing Spatial Plannings
(Planning Evaluation) is to evaluate the implementation of planning
objectives, scale structure, and protection and utilization of the
existing spatial plannings, identify the conflicts and contradictions
between different spatial plannings, and summarize the achievements
and problems.

参考文献/References

［1］国土人. 什么是"双评估"［EB/OL］.（2020-07-12）［2022-10-16］.
https://www. guoturen. com/wenti-1342. html.

术语 4："一张图"

Terminology 4："One Map"

内　涵

"一张图"是指依托国土空间基础信息平台，以一张底图为基础，整合叠加各级各类国土空间规划成果，实现各类空间管控要素精准落地，形成的覆盖全国、动态更新、权威统一的全国国土空间规划"一张图"，可层层叠加打开，为统一国土空间用途管制、实施建设项目规划许可、强化规划实施监督提供依据和支撑[1]。

CONNOTATION

"One Map" refers to national territorial spatial planning "one map" that covers the whole country, is dynamically updated, and is authoritative and unified, relying on territorial spatial basic information platform, based on a base map and the integration and superposition of the outcomes from territorial spatial plannings. It can be opened layer by layer, provide the basis and supports for unifying the national territorial spatial use regulation, implementing planning permission for construction projects and strengthening supervision over the planning implementation.

参考文献/**References**

[1] 中华人民共和国自然资源部. 自然资源部办公厅关于开展国土空间
规划"一张图"建设和现状评估工作的通知 [EB/OL]. (2019-07-18)
[2022-10-25]. https://gi. mnr. gov. cn/202111/t20211129_2708446. html.

术语 5："三区三线"

Terminology 5："Three-Spaces and Three-Lines"

内　涵

"三区三线"是三类空间和三条控制线的简称[1]。

"三区"是指城镇空间、农业空间和生态空间[1]。城镇空间指以承载城镇经济、社会、政治、文化、生态等要素为主的功能空间。农业空间指以农业生产、农村生活为主的功能空间。生态空间指以提供生态服务或生态产品为主的功能空间[2]。

"三线"是指城镇开发边界、永久基本农田和生态保护红线三条控制线。城镇开发边界是在一定时期内因城镇发展需要，可以集中进行城镇开发建设、以城镇功能为主的区域边界，涉及城市、建制镇以及各类开发区等。永久基本农田是为保障国家粮食安全和重要农产品供给，实施永久特殊保护的耕地。生态保护红线是在生态空间范围内具有特殊重要生态功能、必须强制性严格保护的区域[3]。

CONNOTATION

"Three-Spaces and Three-Lines" is the abbreviation of three types of spaces and three control lines.

"Three-Spaces" is Urban Space, Agricultural Space and Ecological Space. Urban Space refers to the functional space which mainly carries urban economic, social, political, cultural, ecological and other factors. Agricultural Space refers to the functional space mainly having agricultural production and rural life. Ecological Space refers to the functional space mainly providing ecological services or ecological products.

"Three-Lines" is three control lines of Urban Development Boundary, Permanent Basic Farmland and Ecological Conservation Redline. Urban Development Boundary is the regional boundary with urban functions as the main function where urban development and construction can be concentrated because of the needs of urban development in a certain period, including cities, towns and various development zones. Permanent Basic Farmland is the cultivated land under permanent special protection to ensure national food security and the supply of important agricultural products. Ecological Conservation Redline refers to the area with special and important ecological functions within the ecological space, which must be strictly protected.

参考文献/References

［1］中华人民共和国中央人民政府. 中共中央办公厅 国务院办公厅印发《省级空间规划试点方案》［EB/OL］. （2017-01-09）［2021-12-28］. http://www. gov. cn/zhengce/2017-01/09/content_5158211. htm.

［2］中华人民共和国自然资源部. 省级国土空间规划编制指南（试行）［EB/OL］. （2020-01-17）［2022-10-02］. http://gi. mnr. gov. cn/2020 01/P020200120642346540184. pdf.

［3］新华社. 中共中央办公厅 国务院办公厅印发《关于在国土空间规划中统筹划定落实三条控制线的指导意见》［EB/OL］. （2019-11-01）［2021-12-17］. http://www. gov. cn/zhengce/2019-11/01/content_5447654. htm.

术语 6："两空间内部一红线"

Terminology 6："Two Spaces and One Redline"

内　涵

"两空间内部一红线"是指将海洋国土空间划分为海洋生态空间和海洋开发利用空间，在海洋生态空间内划定海洋生态保护红线，在海洋开发利用空间内适度留白[1]。

CONNOTATION

"Two Spaces and One Redline" means to delimit marine territorial space into Marine Ecological Space and Marine Development Space, delimit Marine Ecological Conservation Redline in the Marine Ecological Space, and leave blank appropriately in Marine Development Space.

参考文献/References

[1] 自然资源部办公厅. 自然资源部办公厅关于落实海洋"两空间内部一红线"及开展相关试点工作的函，自然资办函[2020]1285 号.

术语7："五级三类四体系"

Terminology 7："Five-Levels, Three-Types and Four-Subsystems"

内 涵

"五级"是指国土空间规划从纵向上对应我国的行政管理体系，分五个层级，就是国家级、省级、市级、县级、乡镇级。其中国家级规划侧重战略性，省级规划侧重协调性，市县级和乡镇级规划侧重实施性[1]。

"三类"是指规划的类型，分为总体规划、详细规划、相关的专项规划[1]。

"四体系"是指规划编制审批体系、规划实施监督体系、法规政策体系和技术标准体系。从规划运行方面，国土空间规划体系分为四个子体系：按照规划流程可以分成规划编制审批体系、规划实施监督体系；从支撑规划运行角度有两个技术性体系，一是法规政策体系，二是技术标准体系。这四个子体系共同构成国土空间规划体系[1]。

CONNOTATION

"Five-Levels" means that Territorial Spatial Planning is vertically corresponding to China's administrative management, which is divided into five levels, namely national, provincial, municipal, county and village level. National level planning focuses on strategy, provincial level plannings focus on coordination, and municipal, county and village level plannings focus on implementation.

"Three-Types" refers to the types of Territorial Spatial Planning, which are divided into Overall Planning, Detailed Planning and relevant Special Planning.

"Four-Subsystems" refers to Planning Formulation and Approval System, Planning Implementation and Supervision System, Laws and Regulations System, and Technical Standards System. From the perspective of planning operating, Territorial Spatial Planning system can be divided into "Four-Subsystems". According to the planning process, it can be divided into Planning Formulation and Approval System, and Planning Implementation and Supervision System. From the perspective of supporting planning operation, there are two technical systems, one is Laws and Regulations System, another is Technical Standards System. These Four Sub-systems constitute Territorial Spatial Planning system.

参考文献/References

[1] 中华人民共和国国务院新闻办公室. 国土空间规划按层级和内容分为"五级三类"[EB/OL]. (2019-05-27) [2021-12-22]. http://www.scio.gov.cn/xwfbh/xwbfbh/wqfbh/39595/40528/zy40532/Document/1655483/1655483.htm.

术语 8：海岸建筑退缩线制度

Terminology 8：Coastal Construction Retreat Line System

内　涵

海岸建筑退缩线制度指海岸建筑退缩线划定与管理工作[1]。海岸建筑退缩线是根据海岸线类型及环境特征，综合考虑海洋灾害、生态环境、亲海空间等要素，基于海岸线向陆一侧延伸一定的距离，划定的禁止或限制建筑活动的控制界线[2]。

CONNOTATION

Coastal Construction Retreat Line System refers to the delimitation and management of Coastal Construction Retreat Line. Coastal Construction Retreat Line is the control lines that is delimited to prohibit or restrict construction activities from the coastline extending a certain distance to the landside, according to the type and environmental characteristics of the coastline, taking into account marine disasters, ecological environment, sea-enjoyable space and other factors.

参考文献/References

[1] 山东省自然资源厅. 关于建立实施山东省海岸建筑退缩线制度的通知[EB/OL].（2022-01-10）[2022-10-14]. http://dnr. shandong. gov.

cn/zwgk_324/xxgkml/ywdt/tzgg_29303/202202/t20220221_3861289. html.

［2］山东省自然资源厅. 山东省海岸建筑退缩线划定技术指南［EB/OL］.
（2022-01-10）［2022-10-14］. http：//dnr. shandong. gov. cn/zwgk_324/
xxgkml/ywdt/tzgg_29303/202202/t20220221_3861289. html.

术语9：自然岸线

Terminology 9：Natural Coastline

内 涵

　　自然岸线是指由海陆相互作用形成的海岸线，包括砂质岸线、淤泥质岸线、基岩岸线、生物岸线等原生岸线。整治修复后具有自然海岸形态特征和生态功能的海岸线纳入自然岸线管控目标[1]。

CONNOTATION

Natural Coastline refers to the coastline formed by the interaction of sea and land，including sandy coastline，muddy coastline，bedrock coastline，biological coastline and other original coastlines. After restoration and rehabilitation，the coastlines with natural coastal morphological characteristics and ecological functions are included in the natural coastline reservation objective.

参考文献/References

［1］中华人民共和国自然资源部. 海岸线保护与利用管理办法［EB/OL］. （2017-03-31）［2021-12-30］. http://gc. mnr. gov. cn/201806/t20180614_1795724. html.

术语 10：自然岸线保有率

Terminology 10：Reservation Rate of Natural Mainland Coastline

内　涵

自然海岸线保有率[1]是指自然海岸线保有量（长度）占海岸线总长度的百分比值[2]。

CONNOTATION

The Reservation Rate of Natural Mainland Coastline refers to the percentage of the natural coastline length in the total length of the coastline.

参考文献/References

[1] 中华人民共和国自然资源部. 海岸线保护与利用管理办法[EB/OL].（2017-03-31）[2021-12-30]. http://gc. mnr. gov. cn/201806/t2018 0614_1795724. html.

[2] 山东省海洋局. 海岸线调查技术规范：DB 37/T 3588—2019[S]. 济南：山东省市场监督管理局，2019.

术语 11：人工岸线

Terminology 11：Artificial Coastline

内 涵

人工岸线是指由永久性人工构筑物组成的岸线[1]。

CONNOTATION

Artificial Coastline refers to the coastline composed of permanent artificial structures.

参考文献/References

[1] 山东省海洋局. 海岸线调查技术规范：DB 37/T 3588—2019[S]. 济南：山东省市场监督管理局，2019.

Part 6

六

国际海洋空间规划

术语 1：欧盟海洋空间规划

Terminology 1：Maritime Spatial Planning in Europe（MSP in Europe）

内　涵

　　欧盟海洋空间规划指令将海洋空间规划定义为：相关成员国当局分析和组织海域人类活动，以实现生态、经济和社会目标的过程。在欧洲，海洋空间规划是欧盟综合海洋政策的一部分，它的目标是支持海洋的可持续发展，为欧盟海洋、岛屿、沿海和最外围地区和海洋行业等相关政策的制定提供协调、连贯和透明的决策支持。

CONNOTATION

　　The EU Directive on MSP defines MSP as：a process by which the relevant Member State's authorities analyse and organize human activities in marine areas to achieve ecological, economic and social objectives. In Europe, MSP is part of the overarching Integrated Maritime Policy（IMP）of the EU which has, as its objective, to "support the sustainable development of seas and oceans and to develop coordinated, coherent and transparent decision-making in relation to the European Union's sectoral policies affecting the oceans, seas, islands, coastal and outermost regions and maritime sector…"[1].

参考文献/**References**

［1］European MSP Platform, European Commission. MSP in Europe ［EB/
OL］. ［2023-05-01］. https://maritime-spatial-planning. ec. europa. eu/
msp-eu/introduction-msp.

术语 2：欧盟海洋空间规划指令

Terminology 2：The EU Directive on MSP

内　涵

该指令为海洋空间规划建立了一个框架，旨在促进海洋经济的可持续增长、海洋区域的可持续发展和海洋资源的可持续利用。

CONNOTATION

This Directive establishes a framework for maritime spatial planning aimed at promoting the sustainable growth of maritime economies, the sustainable development of marine areas and the sustainable use of marine resources [1].

参考文献/References

[1] EU-Lex. MSP in Europe. Directive 2014/89/EU of the European Parliament and of the Council of 23 July 2014 [EB/OL]. (2014-08-01) [2023-02-21]. https://maritime-spatial-planning. ec. europa. eu/msp-eu/introduction-msp.

术语 3：德国海洋空间计划

Terminology 3：Germany's Maritime Spatial Plan(s)

内 涵

德国于 2009 年通过了第一个北海和波罗的海德国专属经济区(EEZ)海洋空间计划。德国的第二个海洋空间计划于 2021 年 9 月生效，涵盖北海和波罗的海的德国专属经济区(EEZ)，以及下萨克森州、石勒苏益格-荷尔斯泰因州和梅克伦堡-前波美拉尼亚州三个沿海联邦州管辖的领海区域。

CONNOTATION

Germany adopted its first maritime spatial plan in 2009 for the German Exclusive Economic Zone (EEZ) of the North Sea and the Baltic Sea. Germany's second maritime spatial plan entered into force in September 2021 and covers the German Exclusive Economic Zone (EEZ) of the North Sea and the Baltic Sea, and the territorial sea areas under jurisdiction of the three coastal federal States：Lower Saxony, Schleswig-Holstein, and Mecklenburg-Vorpommern [1].

参考文献/References

[1] European MSP Platform, European Commission. Germany-Maritime Spatial Plan(s) [EB/OL]. [2023-05-08]. https://maritime-spatial-planning. ec. europa. eu/countries/germany.

术语 4：北海比利时区域海洋空间计划

Terminology 4：Marine Spatial Plan for the Belgian Part of the North Sea

内 涵

北海比利时区域海洋空间计划为比利时领海和专属经济区管理制定了原则、目标、长期愿景和空间政策选择。涉及海洋保护区和包括商业捕鱼、近海水产养殖、近海可再生能源、航运、疏浚、采砂、管道电缆、军事活动、旅游娱乐以及科学研究在内的人类活动的管理行动、指标和目标都包括在内。比利时第一个海洋空间计划是由联邦公共卫生部海洋环境服务局指定的，期限为 2014—2020 年。当前的海洋空间计划期限为 2020—2026 年。

CONNOTATION

The Marine Spatial Plan for the Belgian Part of the North Sea lays out principles, goals, objectives, and long-term vision, and spatial policy choices for the management of the Belgian territorial sea and EEZ[1]. Management actions, indicators and targets addressing marine protected areas and the management of human uses including commercial fishing, offshore aquaculture, offshore renewable energy, shipping, dredging, sand and gravel extraction, pipelines and cables, military activities, tourism and recreation, and scientific research are included[1]. The first Marine Spatial

Plan was drawn up by the Marine Environment Service of the FPS Health for the period 2014—2020. The current Marine Spatial Plan covers the period 2020—2026 [2].

参考文献/References

[1] European MSP Platform, European Commission. A marine spatial plan for the Belgian Part of the North Sea [EB/OL]. [2024-04-25]. https://maritime-spatial-planning. ec. europa. eu/practices/marine-spatial-plan-belgian-part-north-sea.

[2] FPS Health, Food Chain Safety, and Environment. Marine spatial plan [EB/OL]. (2024-04-25) [2024-04-27]. https://www. health. belgium. be/en/marine-spatial-plan.

术语 5：苏格兰国家海洋计划

Terminology 5：Scotland's National Marine Plan

内 涵

苏格兰国家海洋计划于 2015 年发布。该计划涵盖苏格兰近岸水域（12 海里以内）和离岸水域（12 至 200 海里）。针对海洋环境提出清洁、健康、安全、多产和多样化海洋的愿景，以满足自然和人民的长期需要。该计划为苏格兰 200 海里范围内海洋资源的可持续发展制定了战略政策。

CONNOTATION

The Scotland's National Marine Plan was published in 2015. The plan covers both Scottish inshore waters（out to 12 nautical miles）and offshore waters（12 to 200 nautical miles）. The vision for the marine environment is clean, healthy, safe, productive and diverse seas, managed to meet the long-term needs of nature and people. It sets out strategic policies for the sustainable development of Scotland's marine resources out to 200 nautical miles. [1]

参考文献/References

[1] Scottish Government. Scotland's National Marine Plan [EB/OL]. (2015-03-27) [2023-05-30]. https://www. gov. scot/publications/scotlands-national-marine-plan/.

术语 6：北爱尔兰海洋计划

Terminology 6：Marine Plan for Northern Ireland

内　涵

　　北爱尔兰海洋计划将为其海洋区域的监管、管理、使用和保护提供信息和指导。该文件由两个计划组成，一个用于近岸区，一个用于离岸区。北爱尔兰海洋计划草案在 2018 年 4 月到 6 月进行了公众咨询，农业、环境和农村事务部（DAERA）于 2022 年 10 月发布了修订后的公众参与声明（SPP）。公众参与声明最初于 2012 年发布，标志着北爱尔兰海洋规划进程的开始。北爱尔兰海洋计划草案已发布，其愿景是建立一个可持续管理的健康的海洋区域，以促进今世后代的经济、环境和社会繁荣。北爱尔兰海域由近岸区和离岸区组成。近岸区从平均大潮高潮线向外延伸，最多 12 海里，包括潮汐河和狭长海湾。离岸区是指从 12 海里领海界限向东南延伸至北爱尔兰海域外边界（最远距离近岸边界 31 海里）。

CONNOTATION

The Marine Plan for Northern Ireland will inform and guide the regulation, management, use and protection of its marine area. It is a single document made up of two plans, one for the inshore region and one for the offshore region. Public consultation on the draft Marine Plan took place from April to June 2018, DAERA has published a revised Statement of Public Participation (SPP), October 2022. The SPP was originally published in 2012 and marked the beginning of the marine planning process in Northern Ireland [1]. The vision is a healthy marine area which is managed sustainably for the economic, environmental and social prosperity of present and future generations. The Northern Ireland marine area is made up of an inshore and an offshore region. The inshore region extends from the Mean High Water Spring Tide mark out to, at most, 12 nautical miles (nm) and includes tidal rivers and sea loughs. The offshore region is the area that extends south-eastwardly from the 12nm territorial limit to the outer boundary of the Northern Ireland marine area (31nm from the inshore boundary, at its farthest extent). [2]

参考文献/References

[1] Department of Agriculture, Environment and Rural Affairs. Marine Plan for Northern Ireland [EB/OL]. [2023-03-30]. https://www. daera-ni. gov. uk/articles/marine-plan-northern-ireland.

[2] Department of Agriculture, Environment and Rural Affairs. Draft Marine Plan for Northern Ireland [M]. UK: Department of Agriculture, Environment and Rural Affairs, 2018: 8.

术语7：威尔士国家海洋计划

Terminology 7：Welsh National Marine Plan

内　涵

首个威尔士国家海洋计划于2019年11月发布，制定了未来20年海洋可持续利用政策。该计划的愿景是"清洁、健康、安全、多产和生物多样性"的威尔士海洋。通过生态系统方法，自然资源得到可持续管理，海洋健康且具有韧性，支持可持续和繁荣的经济；通过接触、了解和享受海洋环境和海洋文化遗产，提升健康和福祉；通过蓝色增长，创造更多就业机会和财富，帮助沿海社区更具韧性、繁荣和公平，拥有充满活力的文化；通过负责任地部署低碳技术，威尔士海洋地区正为能源安全和气候变化排放目标做出巨大贡献。威尔士海洋计划区域包括约32 000平方千米的海洋面积，其中海岸线为2 120千米。该计划涵盖威尔士近岸区（从平均大潮高潮线到12海里）和离岸区（超过12海里）。

CONNOTATION

The first Welsh National Marine Plan was published on November 2019, setting out their policy for the next 20 years for the sustainable use of our seas. The vision is that Welsh seas are clean, healthy, safe, productive and biologically diverse：Through an eco-system approach, natural resources are sustainably managed and our seas are healthy and resilient, supporting a sustainable and thriving economy; Through access to, understanding of and enjoyment of the marine environment and maritime cultural heritage, health and well-being are improving; Through blue growth more jobs and wealth are being created and are helping coastal communities become more re-silient, prosperous and equitable with a vibrant culture, and; Through the responsible deployment of low carbon technologies, the Welsh marine area is making a strong contribution to energy security and climate change emissions targets. The Welsh marine plan area consists of around 32 000 km^2 of sea, with 2 120 km of coastline. This Plan covers both the Welsh inshore region (from mean high water spring tides out to 12 nautical miles) and offshore region (beyond 12 nautical miles). [1]

参考文献/References

[1] Welsh Government. Welsh National Marine Plan [M]. UK: Crown copyright, 2019: 4.

术语 8：（美国罗得岛州）海洋特别区域管理计划

Terminology 8: Ocean Special Area Management Plan (Ocean SAMP)

内　涵

（美国罗得岛州）海洋特别区域管理计划是海岸资源管理委员会在该计划区域实施管理、规划和适应性管理的工具。海岸资源管理委员会是一个具有监管职能的管理机构。其主要职责是通过实施综合全面的海岸管理计划和向在该州海岸带进行工作的项目发放许可证的方式，保护、开发和修复该州的沿海区域。

CONNOTATION

Ocean Special Area Management Plan (Ocean SAMP) is the regulatory, planning and adaptive management tool that Coastal Resources Management Council (CRMC) is applying to uphold these regulatory responsibilities in the Ocean SAMP study area[1]. The CRMC is a management agency with regulatory functions. Its primary responsibility is for the preservation, protection, development and where possible the restoration of the coastal areas of the state via the implementation of its integrated and comprehensive coastal management plans and the issuance of permits for work with the coastal zone of the state[2].

参考文献/**References**

［1］ Rhode Island Coastal Resources Management Council. Ocean SAMP Volume 1 ［EB/OL］.（2010-10-19）［2023-01-09］. https://seagrant. gso. uri. edu/oceansamp/documents. html.

［2］ Rhode Island Coastal Resources Management Council. About the CRMC ［EB/OL］.［2023-01-09］. http://www. crmc. ri. gov/aboutcrmc. html.

术语9：（加拿大）海洋空间规划区域

Terminology 9：Marine Spatial Planning Areas

内　涵

　　加拿大有5个海洋空间规划区域：不列颠哥伦比亚省南部，太平洋北海岸，纽芬兰和拉布拉多大陆架，圣劳伦斯湾及河口，斯科舍大陆架和芬迪湾。每个规划区都具有独特的生态、社会和文化特征。规划区的边界是根据海洋环境的生物和物理特征以及人类活动水平确定的。这些边界还反映了对海洋空间规划单元可管理范围的考虑，在加拿大海洋空间规划仍是一个相对较新的概念。

CONNOTATION

There are 5 marine spatial planning areas across Canada：Southern British Columbia，Pacific North Coast，Newfoundland and Labrador Shelves，Estuary and Gulf of St. Lawrence，Scotian Shelf and Bay of Fundy. Each planning area is ecologically，socially，and culturally unique. Boundaries for the planning areas were determined by taking into account biological and physical characteristics of the ocean environment and the level of human activity. These boundaries also reflect consideration around what is manageable for the marine spatial planning unit，recognizing that MSP is still a relatively new concept in Canada.[1]

参考文献/**References**

［1］Government of Canada. Marine spatial planning areas ［EB/OL］.（2024-03-21）［2024-03-25］. https：//www. dfo-mpo. gc. ca/oceans/planning-planification/areas-aires/index-eng. html.

术语 10: (澳大利亚)大堡礁海洋公园
分区计划

Terminology 10: Great Barrier Reef Marine Park Zoning Plan

内 涵

(澳大利亚)大堡礁海洋公园是一个多用途区域。2003 大堡礁海洋公园分区计划提供了一系列生态可持续的娱乐、商业和研究机会，并延续了传统活动。分区有助于管理和保护海洋公园的亲海价值。各区域对于允许的活动、禁止的活动和需要许可的活动制定了不同的规则。一些区域还可能限制某些活动的开展方式。

CONNOTATION

The Great Barrier Reef Marine Park is a multiple-use area. The Great Barrier Reef Marine Park Zoning Plan 2003 provides for a range of ecologically sustainable recreational, commercial and research opportunities and for the continuation of traditional activities[1]. Zoning helps to manage and protect the values of the Marine Park that people enjoy. Each zone has different rules for the activities that are allowed, the activities that are prohibited, and the activities that require a permit. Zones may also place restrictions on how some activities are conducted[2].

参考文献/**References**

［1］Australian Government Great Barrier Reef Marine Park Authority. HOW MARINE PARK ZONING WORKS ［EB/OL］.（2024-02-06）［2024-02-13］. https：//www2. gbrmpa. gov. au/access/zoning/how-it-works.

［2］Australian Government Great Barrier Reef Marine Park Authority. Zoning ［EB/OL］.［2024-02-13］. https：//www2. gbrmpa. gov. au/access/zoning.